P9-BXX-519

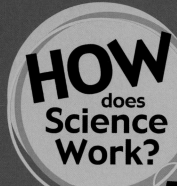

HOW does Science Work?

Exploring Magnets and Springs

Carol Ballard

PowerKiDS press.
New York

Published in 2008 by The Rosen Publishing Group, Inc.
29 East 21st Street, New York, NY 10010

First Edition

Commissioning Editor: Vicky Brooker
Editors: Laura Milne, Camilla Lloyd
Senior Design Manager: Rosamund Saunders
Design and artwork: Peta Phipps
Commissioned Photography: Philip Wilkins
Consultant: Dr Peter Burrows
Series Consultant: Sally Hewitt

Library of Congress Cataloging-in-Publication Data

Ballard, Carol.
   Exploring magnets and springs / Carol Ballard.
       p. cm. — (How does science work?)
   Includes index.
   ISBN 978-1-4042-4282-1 (library binding)
   1. Magnets—Juvenile literature. 2. Springs
(Mechanism)—Juvenile literature. I. Title.

   QC757.5.B365 2008
   538'.4—dc22
                                    2007032266

Manufactured in China

Acknowledgements:

Cover photograph: Spring, Colin Cuthbert/Science Photo Library

Photo credits: Richard Megna/Fundamental/Science Photo Library 4, Michael S. Yamishita/Corbis 6, Charles Gullung/Getty Images 10, Cordelia Molloy/Science Photo Library 14, Tom Van Sant/Geosphere/Corbis 16, Roy Mehta/Getty Images 20,Foodfolio/Alamy 21, PE Reed/Getty Images 22, Colin Cuthbert/Science Photo Library 24, Brad Mitchell/Alamy 26.

The author and publisher would like to thank the models Alex Babatola, Sabiha Tasnim, Zarina Collins, Sophie Campbell, and Philippa Campbell, and Moorfield School for the loan of equipment.

Contents

Words in **bold** can be found in the glossary on p.30

Magnets

Magnets have special **properties**. They can **attract** each other, so that they **pull** toward each other. They can also **repel** each other, so that they **push** away from each other. Magnets can also attract some other materials to them.

Magnets have two ends, each end is called a **pole**. One end is the north pole and the other is the south pole.

Most magnets are made from iron, which is a metal. Magnets come in different shapes and sizes.

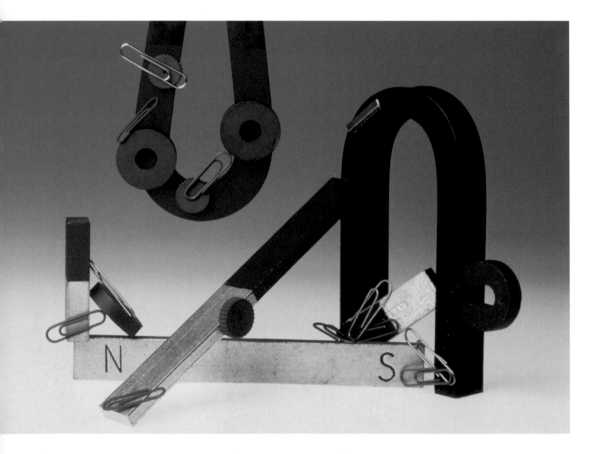

TRY THIS! Investigate magnets

1 You need two bar magnets, the ones used here have one blue end and one red end each.

2 Hold one in each hand.

3 Slowly bring the red ends closer to each other. What can you feel?

4 Now slowly bring the two blue ends closer to each other.

5 Does it feel the same or different?

6 Now bring one red end and one blue end together. How does that feel?

You should find that the blue and red ends are attracted to each other. Two blue ends or two red ends will repel each other.

Comparing magnets

Some magnets are stronger than others, and it is not always the biggest magnet that is the strongest.

Different magnets are good for different jobs. Strong magnets are needed to attract heavy objects, such as old cars and pieces of machinery. Weaker magnets can be used to attract lighter objects, such as pins and paper clips.

This strong magnet attracts metal from other materials in order to collect the metal and recycle it.

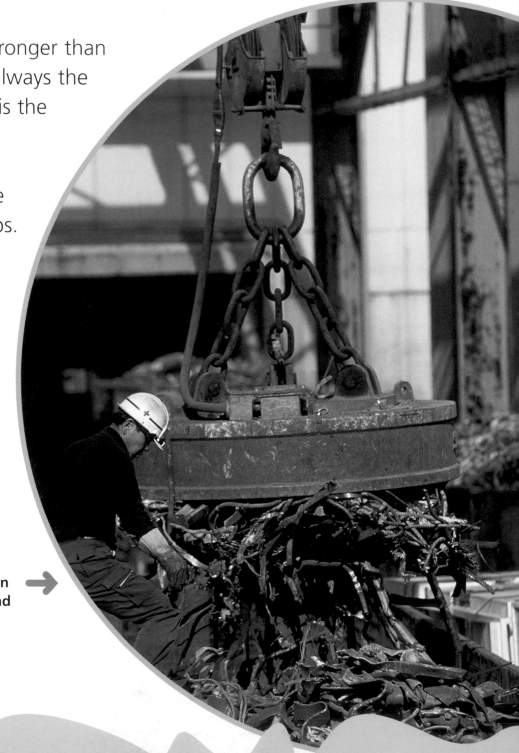

TRY THIS! Test the strength of magnets

1 Find five different magnets.

2 You could choose magnets of different shapes, such as a ball, horseshoe, bar, disk, and strip.

3 Put your first magnet by the 0 in. (0 cm) mark on a ruler.

4 Put a metal paper clip at the 8 in. (20 cm) mark.

5 Slide the paper clip 1/2 in. (1 cm) toward the magnet. Let go of the paper clip.

6 Does the paper clip move toward the magnet?

You should find that some magnets attract the paper clip from farther away than others. These are the strongest magnets.

Is it magnetic?

Something that is attracted to a magnet is called a **magnetic material**. Not all metals are magnetic. Only the metals iron, nickel, and cobalt are magnetic. Iron is a common metal and is used to make many things, such as paper clips, fridge doors, and some parts of cars. Iron is often mixed with other metals. Cobalt and nickel are much less common.

All other materials are **nonmagnetic**. Magnets do not repel nonmagnetic materials—they just have no effect on them at all.

These metal chips contain iron and so they are magnetic.

TRY THIS! Find magnetic materials

1 Choose ten objects around you that are made from different materials.

2 Slowly bring a magnet toward the object to see whether it is attracted to it.

3 Test the other objects one by one.

4 Divide the materials into two groups—a magnetic group and a nonmagnetic group.

You should find that only objects made from metals are magnetic. These must contain some iron, cobalt, or nickel. You might also find that some of your metals are nonmagnetic. These will not contain any of those three metals.

Using magnets

Magnets have many different uses. In the home, some door catches use magnets. You can stick small magnets onto metal surfaces such as fridges for decoration, to hold lists, and to help you to learn. Strong magnets are found in factories and recycling centers. They can be used to sort magnetic materials from nonmagnetic materials, and to pick up heavy things made from magnetic materials.

Electric motors, microphones, loudspeakers, telephones, and video recorders all use magnets to work!

Magnetic letters will stick to metal surfaces such as fridge doors.

TRY THIS! A fishing game

1 You will need a magnet, a stick, and some string for each player, a pile of metal paper clips and some card.

2 Draw lots of fish outlines on your card and cut them out.

3 Attach a paper clip to one side of each fish.

4 Tie a magnet onto one end of the string.

5 Tie the other end onto a stick. Make a fishing rod for each player.

6 With your fishing rod, try fishing for your magnetic fish.

! Be careful using scissors

You should find that your magnet attracts the paper clips attached to your fish. The magnets pick up the magnetic fish.

Magnetism

Inside a magnet there are millions of tiny pieces. Each piece behaves like a minimagnet. They are all lined up so that all the north poles point one way and the south poles point the opposite way.

Magnetic materials contain minimagnets too, but they are mixed up. When you put a magnet close to the magnetic material, the minimagnets line up. We say the material is **magnetized**.

In a **temporary magnet**, the minimagnets slowly get mixed up again and the magnet loses its **magnetism**. In a **permanent magnet**, they stay lined up and magnetic.

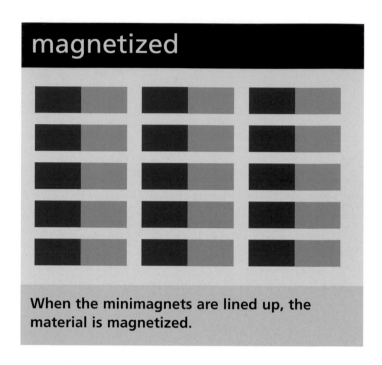

magnetized

When the minimagnets are lined up, the material is magnetized.

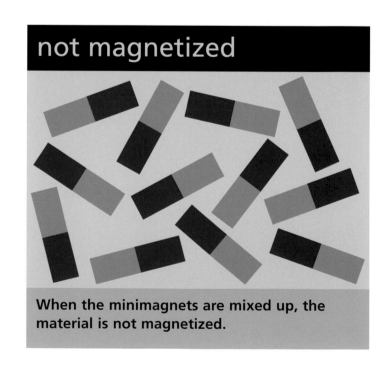

not magnetized

When the minimagnets are mixed up, the material is not magnetized.

TRY THIS! Make a magnet

1 You will need a metal paper clip and a magnet.

2 Gently stroke it from top to bottom with a magnet 50 times in the same direction.

3 Keep your magnet far away from the paper clip between strokes.

4 Your paper clip should now attract other magnetic materials.

You should find that you have made the paper clip into a magnet. You have lined up the minimagnets inside it to make a temporary magnet.

Magnetic fields

Magnets can attract magnetic materials without touching them. They produce a force called magnetism, which works in the space around the magnet.

The **magnetic field** is the space around the magnet. The magnet will attract any magnetic materials inside the magnetic field. Magnetism is strongest around the poles of the magnet and weakest around the center of the magnet.

The iron filings show the pull of the magnetic field around a magnet. →

TRY THIS! Investigate magnetic fields

1 You will need some card, a magnet, and some iron filings.

2 Put a piece of stiff card on top of a magnet.

3 Carefully sprinkle some iron filings on top of the card.

4 Look at the pattern the iron filings make.

The iron filings move because the force of magnetism works through the card. The pattern of the iron filings shows you where the magnetic fields are around the magnet.

You can use adhesive putty or modeling clay to remove the iron filings from the magnets, but make sure you throw it away after use.

! Wash your hands after touching iron filings

Natural magnets

The Earth behaves in the same way as a giant bar magnet. It has its own magnetic field that spreads out into space. The magnetic field is strongest at the poles and weakest around the center.

Hundreds of years ago, people discovered that they could use the Earth's magnetic field to help them find their way. If a magnet is allowed to hang or float freely, the magnet's south pole will point toward the Earth's North Pole. They used this idea to make a simple **compass**, which sailors used to help them find their way. People still use compasses today.

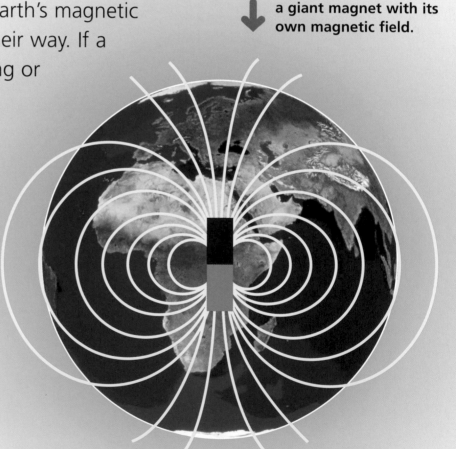

The Earth behaves like a giant magnet with its own magnetic field.

TRY THIS! Make a compass

1   Gently stroke a paper clip with a magnet 50 times, keeping your magnet far away from the paper clip between strokes.

2   Lay the paper clip carefully in a bowl of water so that it floats.

3   Your paper clip will spin around and eventually settle to point in one direction.

4   Hold a compass up to the saucer.

One end of the paper clip will point north and the other will point south. This is because you have made it into a temporary magnet, which will settle with its poles pointing to the Earth's poles.

Springs

Springs are coils that can be squashed or stretched. When you let go, they spring back into their normal shape. Most springs are made from metal. Springs can be long or short, fat or thin. They can be made from thick wire or thin wire.

If the two ends of a spring are pulled, the spring will **stretch**. Some springs are strong and need a big pulling **force** to stretch them. Others are weaker and can be stretched with a smaller pulling force. When a spring is stretched, we say it is under **tension**.

The coils of the springs can be tightly or loosely packed together.

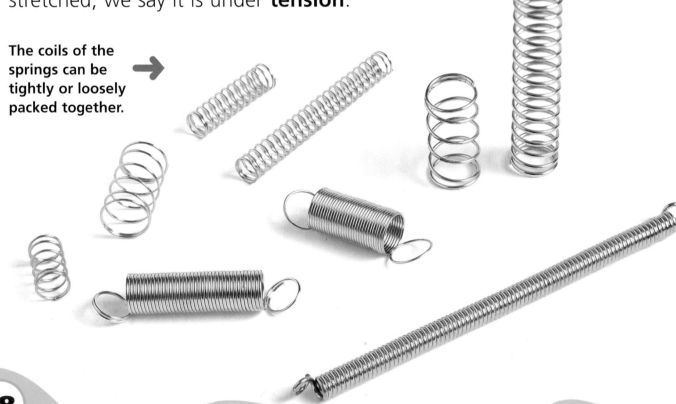

If the two ends of a spring are pushed, the spring will be squashed. Strong springs need a big pushing force to **squash** them. Weaker springs can be squashed with a smaller pushing force. When a spring is squashed, we say it is **compressed**. As soon as you stop pushing on the spring, the spring goes back to its normal shape.

Sometimes, a weak spring can be pulled so hard that it stops behaving like a spring. It just stays stretched and cannot go back to its normal shape.

Some springs can be squashed between your fingers. →

Using springs

Springs are used for many different things. We stretch some springs to use them.

A trampoline has springs around its edges. When you land in the middle, the springs are stretched. As they go back to their normal shape, you are pushed up into the air.

 Springs around this trampoline stretch when you land on it.

We squash some springs to use them. For example, some mattresses have springs inside them. When you lie on them, you squash them. When you get up, they go back to their normal shape. The springs make the mattress feel soft and comfortable.

This scale works because when something is put on it, a spring inside is squashed and makes the pointer move.

Wow!

Train cars, staplers, and push-top pens all have springs in them!

Push and pull

Whatever you do to a spring, it does the same back. To stretch a spring, there must be a pulling force on both ends. These pulling forces must be in opposite directions. When the spring is stretched, it pulls back toward its center. Whatever the size of your pull, the spring pulls back with exactly the same size pull.

To squash a spring, there must be a pushing force on both ends. These pushing forces must be in opposite directions. When the spring is squashed, it pushes back toward its ends. Whatever the size of your push, the spring pushes back with exactly the same size push.

When you push the ends of the clothespin together, the spring is squashed. As you let go, the same size force pushes back to close the clothespin.

TRY THIS! Push and pull a spring

1 Find a spring you can stretch.

2 Gently stretch the spring a little.

3 Can you feel it pulling back?

4 The more you stretch the spring, the more you should feel it pull back.

Now find a spring you can squash. Gently squash the spring. Can you feel it pushing back? If you squash it more, you should feel it pushing back more strongly.

Stretching springs

Springs uncoil when they are stretched. A spring is made from metal wire wound into coils. When a spring is stretched, there is no change in the length of metal or the number of coils. What changes is the distance between the coils.

A **force meter** contains a spring that can be stretched. When you hold the top of the force meter and hang an object on the hook, the object pulls the hook down. This stretches the spring inside. You can read the size of the pulling force by reading the scale on the side of the force meter.

When a spring is stretched, the coils are pulled farther apart.

TRY THIS! Stretch a force meter's spring

1 Measure the length of the spring inside a force meter.

2 Hang a mug or a pencil case on the end, and measure the length of the spring again.

3 Subtract your first measurement from this measurement to find how many inches or centimeters the spring has stretched.

4 Now add a different object and measure the spring again.

Subtract your first measurement again to find out how many inches or centimeters the spring has stretched this time. You should find that the heavier the object, the more the spring will stretch.

You can use a force meter to investigate stretching springs.

Squashing springs

When you squash a spring, you push its coils closer together. The spring will seem shorter, but there is no change in the length of metal or the number of coils.

When the spring is stretched, the coils are pulled farther apart. The spring will look longer but there will still be the same number of coils. There is just more space between each coil.

There is a spring inside this jack-in-the-box. It is squashed when the lid is shut, but pushes the toy up when the lid opens.

TRY THIS! Make a pop-up card

1   You will need some card, a spring, scissors, and some adhesive putty.

2   Fold a piece of card in half.

3   On one side, decorate the front of your card.

4   Make a flower or animal shape from another piece of card.

5   Attach the spring to the animal or flower, and stick it on the inside of the card with putty or tape.

6   Shut the card.

When you open it, the spring should pop up with your shape on the end!

! Be careful using scissors

Stretchy materials

Stretchy materials have similar properties to springs. Some materials will stretch and then go back to their original shape. These materials are **elastic materials**. For example, rubber bands will go back to their original shape after they have been stretched. Other materials, such as modeling clay, can stay in their new shape. Materials such as this are not elastic.

These things are all made from elastic materials.

TRY THIS! Investigate stretchy materials

1   Gather some materials that can be stretched, such as a piece of elastic, a rubber band, an unused balloon, and some old tights.

2   Stretch each material against a ruler.

3   You can make a chart of how far each material stretched.

Which material stretched the most? Which material stretched the least? Did any materials stay stretched, or were they all elastic materials?

You should find that the elastic materials spring back to their original shape.

Glossary

attract pull toward

compass an instrument that tells you which direction you are facing

compressed when a spring is squashed

elastic materials materials that go back to their original shape after being stretched

force a push or a pull

force meter an instrument for measuring forces

magnetic material a material that acts like a magnet

magnetic field the space around a magnet where its magnetic force works

magnetized materials that become magnetic

magnetism the force that pulls magnetic materials toward a magnet

magnets objects that attract magnetic materials to them

nonmagnetic a material that does not act like a magnet, and is not affected by a magnet

permanent magnet a magnet that is always magnetic

pole one end of a magnet

properties what a material is like

pull force used to stretch a spring or stretchy material

push force used to squash a spring or stretchy material

repel push away

springs coils of wire that can be stretched or squashed

squash get shorter

stretch get longer

temporary magnet a magnet that quickly loses its magnetism

tension the pull in a spring or stretchy material when it is stretched

Further information

Books to read

Magnetic and Non-Magnetic (My World of Science) by Angela Royston (Heinemann Library, 2004)

Magnetism: From Pole to Pole by Chris Cooper (Heinemann Library, 2003)

Magnets (Science Experiments) by Sally Nankivell-Aston and Dot Jackson (Franklin Watts Ltd., 2000)

Magnets: Sticking Together by Wendy Sadler (Raintree, 2005)

Web sites to visit

Web Sites
Due to the changing nature of Internet links, PowerKids Press has developed an online list of Web sites related to the subject of this book. This site is regularly updated. Please use this link to access this list:
www.powerkidslinks.com/hdsw/magspr

CD Roms to explore

Eyewitness Encyclopedia of Science, Global Software Publishing

I Love Science!, Global Software Publishing

My First Amazing Science Explorer, Global Software Publishing

Index